Imprint:

Copyright © 2015 GRIN Verlag
Print and binding: Books on Demand GmbH, Norderstedt Germany
ISBN: 9783668833906

This book at GRIN:

https://www.grin.com/document/441517

Florante Jr. Poso

Properties of Fluids in an Engineering Context

GRIN Verlag

PROPERTIES OF FLUIDS: In an engineering context

Florante C. Poso, Jr., MCE, PhD

2018

Table of Contents

Introduction

The eBook discusses the different properties of fluids. In a general context, fluids are classified as either a liquid or gases, although in other textbooks, it also incorporates plasma as part of its scope. The coverage of this eBook is limited to the detail discussion of the fluid properties such as the density, viscosity, compressibility and elasticity, vapor pressure, surface tension and capillary rise or depression. Specific problems are presented with detailed solutions to guide the reader on the step by step procedure of solving in an engineering point of view.

The study of fluid mechanics is utilised in the field of engineering most specifically on engineering structures that incorporate the conveyance of fluids in the system. Pipelines and machines like turbines and engines also use the principles of fluid mechanics. Thus, it is cognizant to learn the basics of the properties of fluids in order to have a better grasp of the principles of fluids in application to engineering works. The units that were used in the illustrations are all in metric system.

Fluids

"Fluid mechanics is a branch of physics concerned with the mechanics of fluids (liquids, gases, and plasmas) and the forces on them."[1]

"Fluids are divided into liquids and gases. A liquid is hard to compress and it changes its shape according to the shape of its container with an upper free surface. Gas on the other hand is easy to compress, and fully expands to fill its container."[2]

"There are two aspects of fluid mechanics which make it different to solid mechanics. Firstly, the nature of fluid is much different to that of a solid, and secondly, in fluids we usually deal with continuous streams of fluid without a beginning or end. In solids we only consider individual elements."[3]

---

[1] Fluid mechanics, https://en.wikipedia.org/wiki/Fluid_mechanics
[2] Y. Nakayama, Introduction to Fluid Mechanics, Butterworth Heinemann, Oxford, 2000, p6
[3] http://site.iugaza.edu.ps/ymogheir/files/2010/02/Chapter1_Fluids_Properties_KA_YM_06Sep15.pptx

Properties of Fluids

The different properties of fluids are discussed below:

Density

> "The density of a fluid, is generally designated by the Greek symbol ρ (rho) is defined as the mass of the fluid over an infinitesimal volume."[4] Is oil lighter than water? This kind of question actually compares two measures, the mass and the volume, which makes sense in the determination of the density of a certain fluid.

Mass Density

> "Mass Density is generally defined as the mass per unit volume of a material."[5]

$$\rho = m / V$$

where:

ρ = density or mass density in kg/m^3
m = mass in kg
V = volume in m^3

Problem 1.

Compute the mass density of oil that has a mass of 835 kg with a volume of 0.92 m^3.

$$\rho = m / V$$
$$\rho = 835 \text{ kg} / 0.92 \text{ m}^3$$
$$\rho = 907.61 \text{ kg} / m^3$$

Specific Weight

> "The specific weight of a fluid is designated by the Greek symbol γ (gamma), and is generally defined as the weight per unit volume."[6]

$$\gamma = \rho g$$

the units used is N/m^3 or kN/m^3
where:

---

[4] Fluid Mechanics/Fluid Properties
https://en.wikibooks.org/wiki/Fluid_Mechanics/Fluid_Properties, 19 October 2017
[5] Ibid.
[6] Ibid.

γ = Specific weight in N/m³ or kN/m³
ρ = density or mass density in kg/m³
g = gravitational acceleration (9.807 m/s²)

For example, calculating the Specific weight of water, given the mass density of water (ρ) as 1000 kg/m³ and the value of g = 9.807 m/s².

$\gamma = \rho \, g$

γ = (1000 kg/m³) (9.807 m/s²)

γ = 9807 N/m³ or 9.807 kN/m³

Calculating the Specific weight of air where the mass density of air (ρ) as 1.205 kg/m³ and the value of g = 9.807 m/s².

$\gamma = \rho \, g$

γ = (1.205 kg/m³) (9.807 m/s²)

γ = 11.82 N/m³

Relative Density

"The relative density is the ratio of the density (mass of a unit volume) of a substance to the density of a given reference material, which is mostly specified as water."[7] Relative density or sometimes called as specific gravity is unit less.

Water is mostly used as reference liquid which has a density of ρ=1000 kg/m³.

Based on the definition, the formula for relative density is:

$$RD = \frac{\rho\,(liquid)}{\rho\,(water)}$$

where:
RD = relative density or SD = specific density

---

[7] Relative density, https://www.chemsafetypro.com/Topics/CRA/Relative_Density.html

Problem 2.

A mass of a container of a given liquid is 800 kg. Given the volume of the container is 0.90 m³. Find the values of the density, specific weight, and relative density or specific gravity of the liquid.

$$\rho \ of \ liquid = \frac{m}{V}$$

$$\rho \ of \ liquid = \frac{800}{0.90} = 888.89 \ kg/m^3$$

$$\gamma \ of \ liquid = \frac{weight}{volume} = \frac{m \ g}{V}$$

$$\gamma \ of \ liquid = \frac{800 \ x \ 9.81}{0.90} = 8720 \ N/m^3$$

$$RD \ of \ liquid = \frac{\rho \ of \ liquid}{\rho \ of \ water}$$

$$RD \ of \ liquid = \frac{800}{1000} = 0.80$$

Viscosity

"Viscosity is the quantity that describes a fluid's resistance to flow."[8] There are two (2) classifications of viscosity, the dynamic and kinematic viscosity.

"The Newton's law shows the relationship between shear stress and viscosity as shown in the equation and figure below:"[9]

$$\tau = \mu \frac{du}{dy}$$

---

[8] Viscosity, The Physics of Hypertextbook, https://physics.info/viscosity/
[9] Zerihun Alemayehu, Hydraulics I, https://aaucivil.files.wordpress.com/2009/10/ch01.pdf

where:

τ = shear stress in N/m^2

μ = coefficient of dynamic viscosity in Ns/m^2

du/dy = velocity of gradient (radian/s)

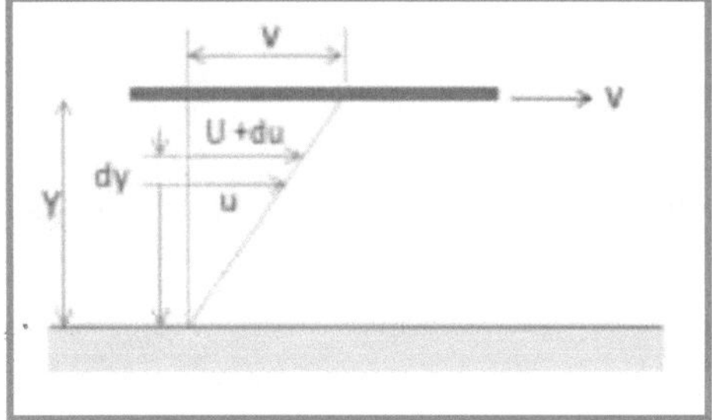

Figure 1. Illustration of Viscosity
Source: (https://aaucivil.files.wordpress.com/2009/10/ch01.pdf)

Dynamic Viscosity (μ)

"The dynamic viscosity of a fluid expresses its resistance to shearing flows, where adjacent layers move parallel to each other with different speeds."[10]

"It is the shear force per unit area required to drag one layer of fluid with unit velocity past another layer a unit distance away."[11]

The unit used for dynamic viscosity is kg/m.s or N.s/m^2

Kinematic Viscosity (v)

"Kinematic viscosity is defined as the ratio of dynamic viscosity to mass density."[12]

From the definition above, equation for kinematic viscosity is:

$$v = \frac{\mu}{\rho}$$

---

[10] Viscosity, https://en.wikipedia.org/wiki/Viscosity#Dynamic_(shear)_viscosity

[11] Zerihun Alemayehu, Hydraulics I, https://aaucivil.files.wordpress.com/2009/10/ch01.pdf

[12] Ibid

where:

v = kinematic viscosity

μ = dynamic viscosity

ρ = mass density

The unit used for kinematic viscosity is m²/s

A certain liquid at 20⁰C has a density of 780 kg/m³. If the dynamic viscosity of the liquid is given as 4.5 x 10⁻³ kg/m.s, calculate its relative density and kinematic viscosity.

Solution:

For relative density (σ)

$$\sigma = \frac{\rho \; of \; liquid}{\rho \; of \; water}$$

$$\sigma = \frac{780}{10^3}$$

$$\underline{\sigma = 0.78}$$

For kinematic viscosity (v)

$$v = \frac{\mu}{\rho}$$

$$v = \frac{4.5 \; x \; 10^{-3}}{780}$$

$$\underline{v = 5.77 \; x \; 10^{-6} \, m^2/s}$$

Compressibility and Elasticity

"The Bulk Modulus of Elasticity or sometimes called as the Volume Modulus is a material property characterizing the compressibility of a fluid."[13]

The Bulk Modulus Elasticity can be calculated as:[14]

$$K = -dp / (dV / V0)$$

$$K = -(p1 - p0) / ((V1 - V0) / V0)$$

where:

K = Bulk Modulus of Elasticity (Pa, N/m^2)

dp = differential change in pressure on the object (Pa, N/m^2)

dV = differential change in volume of the object (m^3)

V_0 = initial volume of the object$_0$ (m^3)

p_0 = initial pressure (Pa, N/m^2)

p_1 = final pressure (Pa, N/m^2)

V_1 = final volume (m^3)

---

[13] Bulk Modulus and Fluid Elasticity, https://www.engineeringtoolbox.com/bulk-modulus-elasticity-d_585.html
[14] Ibid

Problem 4.

A pressure of 3 MPa is applied to a mass of a certain liquid that is initially filled with a volume of 1,500 cm³. Find the volume after the pressure is applied if the Bulk Modulus Elasticity (K) is 2.15 x 10⁹ Pa.

Solution:

$$K = -\, dp \,/\, (dV \,/\, V_0)$$

$$dV = -\frac{dp}{K}\, Vo$$

$$dV = -\frac{3 \times 10^6\ Pa}{2.15 \times 10^9\ Pa}\, 1500\ cm^3$$

$$dV = -2.093\ cm^3$$

$$Vf = Vo + dV$$

$$Vf = 1500 - 2.093\ cm^3$$

$$\underline{Vf = 1497.907\ cm^3}$$

Vapor Pressure

"Vapor pressure is the pressure at which a liquid will boil for a given temperature. Basically, vapor pressure increases with temperature."[15] Contrariwise, if the vapor pressure decreases, the temperature also decreases.

The factors that affect the vapor pressure are the surface area, type of molecules and temperature.

---

[15] Zerihun Alemayehu, <u>Hydraulics I</u>, https://aaucivil.files.wordpress.com/2009/10/ch01.pdf

"Surface tension is the tensile strength per unit length of assumed section on the free surface."[16]

Table 1 shows the surface tension of some liquids.

Table 1. Surface tension at 20⁰C.

Liquid	Surface Liquid	N/m
Water	Air	0.0728
Mercury	Air	0.476
Mercury	Water	0.373
Methyl alcohol	Air	0.023

Source: Y. Nakayama, *Introduction to Fluid Mechanics*, Butterworth Heinemann, Oxford, 2000

Capillary Rise / depression

"Capillarity is a phenomenon that explains the interplay of the forces of cohesion and adhesion. The area of contact between the liquid and solid increases and the liquid thus wets the solid surface."[17]

If the adhesion is greater than cohesion, it is termed as capillary rise, while the other way is called the capillary depression. Figure 2 presents the phenomenon of capillary rise and capillary depression.

---

[16] Y. Nakayama, Introduction to Fluid Mechanics, Butterworth Heinemann, Oxford, 2000, p13
[17] Capillarity, https://nptel.ac.in/courses/112104118/lecture-2/2-9-capilarity-vapour.htm

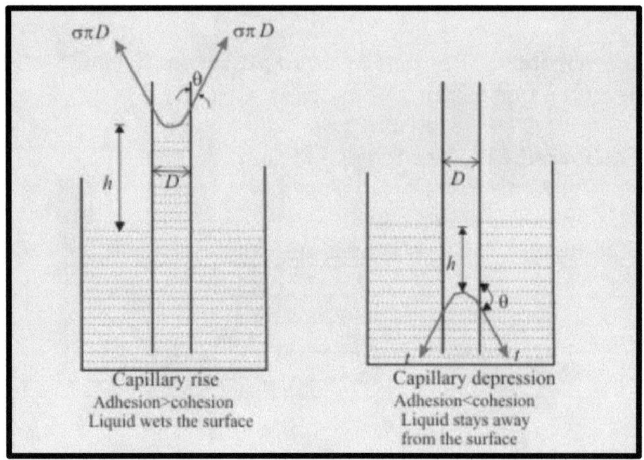

Figure 2. Illustration of Capillary rise and depression
Source: https://nptel.ac.in/courses/112104118/lecture-2/2-9-capilarity-vapour.htm

To determine the Δh in Figure 3 using a capillary tube:[18]

$$\Delta h = \frac{4\sigma\cos\theta}{\gamma d}$$

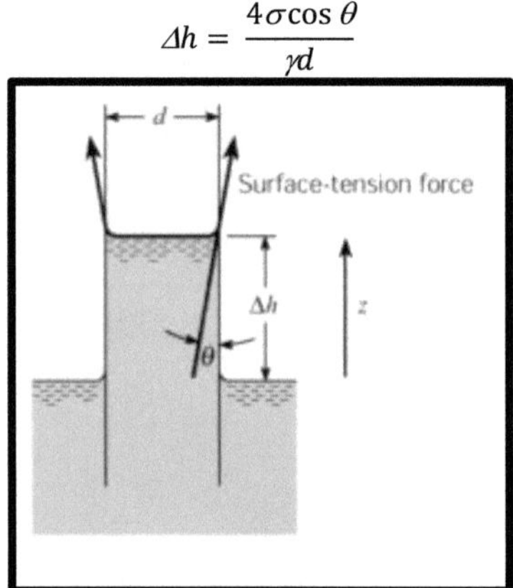

Figure 3. Surface tension
Source: Zerihun Alemayehu, <u>Hydraulics I</u>,

---

[18] Zerihun Alemayehu, <u>Hydraulics I</u>,
https://aaucivil.files.wordpress.com/2009/10/ch01.pdf

Problem 5:

Calculate the wetting angle in a 4 mm diameter vertical tube, if the liquid rises at 7mm above the liquid outside the capillary tube. Assume the surface tension of water to be 0.45 N/m.

$$h = \frac{4\sigma\cos\theta}{\rho g d}$$

Where:

$$\gamma = \rho g = 9810$$

$$d = 0.004 \text{ m}$$

$$h = 0.007 \text{ m}$$

$$\sigma = 0.45 \text{ N/m}$$

$$\therefore \quad \cos\theta = \frac{\gamma d h}{4\sigma} = \frac{9810 \times 0.004 \times 0.007}{4 \times 0.45}$$

$$\cos\theta = 0.1526$$

$$\underline{\theta = 81.22^0}$$

Conclusion:

The study of fluid properties is used throughout the different fields of engineering that are concerned in the conveyance of fluids. In the field of civil engineering, these are the dams and reservoirs, water and wastewater treatment systems, pumping stations, water storage tanks, water pipe systems, waste water collection systems, and storm water conveyance systems and other engineering applications.

Therefore, to understand the full context of the design principles in fluid mechanics, an engineer must have a solid grasp of the properties of fluids first.

References:

Bulk Modulus and Fluid Elasticity, https://www.engineeringtoolbox.com/bulk-modulus-elasticity-d_585.html

Capillarity, https://nptel.ac.in/courses/112104118/lecture-2/2-9-capilarity-vapour.htm

Fluid mechanics, https://en.wikipedia.org/wiki/Fluid_mechanics

http://site.iugaza.edu.ps/ymogheir/files/2010/02/Chapter1_Fluids_Properties_KA_YM_06Sep15.pptx

Relative density, https://www.chemsafetypro.com/Topics/CRA/Relative_Density.html

Viscosity, https://en.wikipedia.org/wiki/Viscosity#Dynamic_(shear)_viscosity

Viscosity, The Physics of Hypertextbook, https://physics.info/viscosity/

Y. Nakayama, Introduction to Fluid Mechanics, Butterworth Heinemann, Oxford, 2000, p13

Zerihun Alemayehu, Hydraulics I, https://aaucivil.files.wordpress.com/2009/10/ch01.pdf